Aerospace Engineering and Virtual Reality

A Next-Generation Partnership

Table of Contents

Chapter 1. Introduction

Delving into the realms both of our earthly confines and virtual illusions, this Special Report presents a comprehensive yet straightforward discussion on the sweeping partnership between Aerospace Engineering and Virtual Reality. Despite the complex technological interplay at work, this document ensures to unpack this riveting convergence in the simplest and most graspable terms for our readers. The pages that follow are brimming with insights into how the intricate mechanisms of aerospace are being simulated and perfected using virtual reality, and how this collaboration is shaping the future of human space exploration. Designed to be both informative and engaging, this Special Report is a must-have for anyone interested in understanding the next wave of technological symbiosis, no matter your level of technical expertise. Lend us your curiosity, and we promise a journey across the vistas of reality and virtuality, all while keeping you firmly grounded.

Chapter 2. The Confluence of Aerospace Engineering and Virtual Reality

The dawn of the 21st century saw technology breaking barriers and crossing thresholds once believed to be untouchable. Two such fields that have made significant strides are Aerospace Engineering and Virtual Reality. While at first glance, the interfusion of these fields may seem abstract, the fusion turns out to be a match more heavenly than terrestrial.

2.1. The Unyielding Flight of Progress

In the heart of our civilizations' progress, there has been a consistent force, engineering. Engineering has allowed us to craft and mold our reality in ways that have outpaced even the wildest imaginations. Aerospace Engineering, in particular, has driven us to the outer reaches of our planet, and eventually, to the stars.

Simultaneously, another realm has taken flight, offering untamed expansion to creativity and exploration: Virtual Reality. VR has provided a new dimension to understand and interact with our environment by generating artificial settings that can be as real as our world or as fantastical as our dreams.

The merging of these radical disciplines, Aerospace Engineering and Virtual Reality, has opened a world of possibilities that promises to redefine the terrain of human exploration.

2.2. Man and the Final Frontier

Space, referred to as the final frontier, has remained the ultimate challenge to human exploratory endeavors. The void beyond our atmosphere is filled with wonders and dangers alike, pushing our engineering technologies to their limits.

Aerospace engineering uses principles of physics and mathematics to design, develop, and apply aircraft and spacecraft. The risks associated with manned space missions demand meticulous calculation and preparation. However, the prohibitive costs, time needed, and inherent dangers make repeated testing nearly impossible.

This is where the confluence of Aerospace Engineering and Virtual Reality becomes more than just a novelty. It offers a lifeline—an opportunity to merge the colossal outreach of space-faring technologies with the risk-free, iterative design and simulated testing environment that VR offers.

2.3. The Virtual Reality Advantage

Virtual Reality capitalizes on computer technology to create a simulated, three-dimensional world that a user can manipulate and explore while feeling as if he were in that world. All this is accomplished by stimulating the sensory perceptions of the user, replicating aspects like vision, hearing, touch, and sometimes even smell.

In Aerospace Engineering, VR brings scalability, safety, and cost-effectiveness to the table. Prototypes of spacecraft or aeronautical vehicles can be created, modified, and tested in a simulated environment. This delivers enormous cost savings as the physical creation of these models is both expensive and time-consuming.

Besides, virtual prototyping permits comprehensive and iterative testing without endangering lives or risking expensive equipment. Critical mission elements, such as vehicle launch, extraterrestrial landing, and even daunting rescue scenarios, could be rigorously simulated and assessed to ensure the maximum success probability.

2.4. Case Studies in Aerospace VR Applications

Notably, organizations like NASA have effectively deployed Virtual Reality in their operations. The Mixed Reality Lab of NASA uses VR and AR (Augmented Reality) tools to support an array of projects. One such application is spacecraft interior design, where designers can virtually "enter" a spacecraft and experience its functionality before it is built.

On the other side of the pond, the European Space Agency (ESA) utilizes VR for extravehicular activity (EVA) training. Astronauts use VR systems to simulate their environment and rehearse their tasks, preparing them for the real challenges they will face in the void of space.

Many private space venture firms, such as SpaceX and Blue Origin, have also integrated VR into their development and testing processes for crewed mission training. Prototypes of Mars rovers, lunar modules, and other spacecraft are virtually designed, manipulated, and tested for high-risk scenarios to ensure safety and mission success.

2.5. Future Trajectories

As we stand on the brink of a new era of exploration—with missions to Mars and the potential colonization of other worlds—the partnership between Aerospace Engineering and Virtual Reality will inevitably deepen. Virtual prototypes will refine designs faster and cheaper, and immersive training environments will prepare astronauts better than any physical simulator. Furthermore, VR might facilitate remote construction of space habitats through teleoperation robots, revolutionizing the way we think about space colonization.

Through a rigorous embrace of Virtual Reality, Aerospace Engineering steps into a new epoch of safer and more efficient exploration. Together, they represent the best of human imagination and ingenuity.

In conclusion, this exciting convergence of Aerospace Engineering and Virtual Reality is a testament to the remarkable pace with which our world is digitizing. This alliance is not just reshaping space exploration practices; it's redefining them. The future of next-level engineering and space discovery lies within the embrace of this powerful duo, perpetually ready to conquer uncharted terrain, whether it resides on our home planet, the endless void of space, or the immersive realm of the virtual world.

Chapter 3. Fundamentals of Aerospace and its Simulation in VR

Aerospace engineering, a field focusing on the design, development, testing and optimization of both aircraft and spacecraft, is one of the most technologically challenging fields of engineering. There's no room for error when creating vehicles that operate in the unforgiving environment of space or the challenging atmosphere of our planet. Virtual reality (VR), offering immersive, interactive, and three-dimensional environments, has turned out to be a powerful tool in this context. It aids in attaining unprecedented precision, efficiency, and safety during the design, testing, and training phases of aerospace engineering ventures.

3.1. The Deeper Dive into Aerospace Engineering

Aerospace engineering is a truly multidisciplinary field, bringing together elements from mechanical engineering, electronics, material science, computer science, and more to create advanced vehicles for air and space travel. Two primary branches constitute Aerospace Engineering - Aeronautical Engineering (concerned with aircraft that operate within the Earth's atmosphere) and Astronautical Engineering (dedicated to the design and development of spacecraft).

Designing and developing aircraft and spacecraft is a highly complex process. It starts with conceptual design, followed by a detailed design phase, then moves to simulation and testing, before finally reaching the manufacturing and production stage. Each stage has its unique set of challenges that need careful resolution.

3.2. VR's Role in Aerospace Engineering

Virtual reality takes the principle of visualizing concepts to a new height in the realm of aerospace engineering. With VR's help, engineers immerse themselves in a virtual environment, where they can interact with a realistic, 3D model of their design. This technology's immersive nature makes it easier for engineers to spot potential design flaws and rectify them early in the design process.

Moreover, it is not just about finding glitches in design but a comprehensive understanding of how different elements of the design interact with each other in a real-world scenario. It's possible to simulate factors such as pressure, temperature, and mechanical stresses that the design will be exposed to in reality — this ability marks a significant leap from traditional ways of aircraft and spacecraft design.

3.3. The Science of Simulation in VR

The core of VR lies in its ability to create convincing simulations of reality. These simulations are created by a combination of 3D computer graphics, real-time game physics, and high-performance computing. The aircraft or spacecraft is modeled in VR down to meticulous detail, making them behave as closely as possible to how they would in real life.

However, the accuracy of this simulation is not solely for visual purposes. VR simulations are used to perform 'virtual tests' that enable engineers to gauge how different aspects of the design will perform in their operating environments but without the associated risks of physical testing. A virtual endurance test to determine how components will fare in extreme temperatures, for instance, can be safely conducted in the simulation without damaging the actual components.

3.4. The Aerospace-VR Collaboration Cycle

The integration between Aerospace Engineering and VR is a cyclic process iterating through design, simulation, testing, and finalization stages. Initially, a design is conceptualized and brought to life in VR. This virtual model undergoes tests in simulated environments to analyze stress points, structural integrity, and possible enhancements. After refining, the enhanced design repeats the test cycles until engineers approve the model. The design then transitions from VR to production.

3.5. Training with VR in Aerospace

Beyond design and testing, VR serves an instrumental role in training scenarios within aerospace engineering. Training programs for astronauts, pilots, and maintenance personnel are now using VR to provide safe, controllable, yet realistic training scenarios.

Astronauts can rehearse complex tasks they'll need to do in space (such as carrying out repairs on the International Space Station). Pilots can practice in realistic flight simulations for various types of aircraft. Complex maintenance tasks can be practiced by ground staff, who can familiarize themselves with intricate aircraft systems and components, all within a VR environment.

3.6. Conclusion

The intersection of VR and Aerospace engineering has challenged and reshaped the boundaries traditionally associated with the design, development, testing, and training processes within the aerospace sector. This technological alliance has facilitated production efficiencies that are financially viable and has contributed to the creation of safer, more efficient aerospace vehicles.

This technological fusion is not just contributing to our endeavors to better traverse our Earth's atmosphere but propelling us safely into the deep expanses of space. This intertwined journey of Aerospace and VR continues, and this report follows its progression, exploring how our human journey, too, is enhanced by it - from Earth to the cosmos, between reality and the virtual.

Chapter 4. Virtual Reality: Transforming Aerospace Training

In the realm of aerospace engineering, the importance of advanced and effective training cannot be overstressed. As technology pushes the boundaries of what humans can achieve, training methodologies must keep pace. Virtual Reality (VR), a technology previously relegated to the gaming industry and nascent tech startups, has surged forward as a catalyst for revolutionizing aerospace training.

4.1. VR: A Simulated Environment

Virtual Reality, in its most fundamental essence, is a technology that allows users to immerse themselves in a simulated environment. With highly sophisticated hardware and software, it crafts an illusion of reality so convincing that our cognitive mechanisms accept it as genuine.

The environments carved by VR are generated by computer graphics that tap into our visual, auditory, and sometimes even tactile senses. The technology allows users to explore and interact with these artificial realities, therefore providing a more enriched and immersive experience compared to traditional forms of media.

4.2. Real-world Applications of VR in Aerospace Training

With the costs associated with aerospace development and its high-stakes nature, VR provides a viable, cost-effective, and safer alternative to real-world training methodologies. Its applications range from design conceptualization to maintenance inspections, pilot training, and mission rehearsals.

One of the primary uses of VR in aerospace training relates to aircraft maintenance and repair. Usually, this requires years of hands-on experience and costly practice parts. With VR, however, trainees from all over the world can practice maintenance procedures in a highly realistic, simulated environment, gathering hands-on experience without risking damage to costly equipment or the safety issues associated with actual aircraft.

4.3. Real-time Mission Planning and Rehearsal

VR technology has also been instrumental in revolutionizing pre-flight preparations and mission rehearsals. As visualization grants a deeper understanding that blueprints or diagrams could never achieve, VR allows astronauts and ground control to practice and improve every significant aspect of a mission, culminating in enhanced proficiency and reduced mistakes.

NASA, for instance, utilizes VR extensively in astronaut training. Their Hybrid Reality Lab uses VR to create highly synchronized and comprehensive space simulations, allowing astronauts to practice operations and familiarize themselves with their work environment before they even leave Earth. This simulated "spacewalk" experience significantly reduces the risks associated with real outer space missions.

4.4. Individual and Team-based Training

VR not only allows individual practice but also enables team-based training. In the aerospace environment, individuals need to work cohesively as a team to ensure mission success. VR provides joint training experiences, fostering a frictionless, collaborative environment regardless of geographical barriers. Haptic feedback provisions, furthermore, allow the incorporation of physical elements, intensifying the robustness of the training process.

4.5. Reducing Costs and Increasing Safety

VR reduces the need for physical resources in aerospace training. This digitalization not only contributes to significant financial savings but also results in fewer environmental implications. Importantly, it drastically reduces the risk of accident or error during training, which is often expensive and sometimes fatal.

4.6. Future of VR in Aerospace Training

VR's current applications in aerospace training are merely the tip of the iceberg in terms of its full potential. As the technology evolves, VR will provide even more immersive and realistic simulations, thus cultivating a generation of aerospace technicians, engineers, and astronauts who can handle complex tasks with unprecedented levels of confidence and competence.

On the horizon are developments such as Augmented Reality (AR) and Mixed Reality (MR), which combine elements of the real and virtual worlds. These technologies will borrow from and improve upon VR, making realistic, high-risk training simulations even more accurate and effective, thus elevating the overall quality of aerospace training.

In conclusion, VR has already begun its journey in transforming aerospace training, making this high risk, high stakes field more accessible and safe. As the technology advances and becomes more widely adopted, the boundary between reality and virtuality in aerospace training will continue to blur, bringing with it an entirely new era of space exploration.

Chapter 5. Case Studies: Aerospace Enterprises Embracing VR

The last few years have seen a revolutionary shift in how aerospace industries operate. Virtual Reality (VR) has become an integral part of training, designing, and testing mechanisms within these enterprises. To better illustrate the underlying dynamics of this transformation, let's delve into the success stories of four major aerospace corporations and how they are leveraging VR.

5.1. The Trailblazer: NASA

NASA, an early adopter of VR, has been using this technology since the 1990s. Today, their application extends to astronaut training, mission simulation, and biological studies, among others. In the Advanced Virtual Environment Training System (AVERTS) developed by NASA, astronauts immerse themselves in a virtual representation of the International Space Station (ISS) for training purposes. This technique allows them to familiarize themselves with the ISS modules and rehearse Extra-Vehicular Activities (EVAs), crucial for their mission success.

NASA also uses VR to visualize space data, which not only aids astronomers' understanding of the cosmos but also fuels public outreach efforts. For instance, the Mars 2030 VR suite offers an interactive journey to the red planet, based on actual NASA data.

5.2. An Odyssey in Innovation: SpaceX

SpaceX has integrated VR into its design and manufacturing process. Engineers use VR to visualize spacecraft components during the design phase, facilitating efficient communication about design changes among team members. This design-test-refine routine enables SpaceX to optimize designs before physical manufacturing, reducing lead times and cost.

In 2019, SpaceX made headlines when it used VR in the critical process of docking the Dragon spacecraft with the ISS. The astronauts for the Crew Dragon Demo-2 mission trained in a highly accurate VR simulation of the spacecraft's cockpit, enhancing their preparedness. This not only made the astronauts comfortable with the controls and procedures but also enabled them to react promptly to any anomalies during the live operation.

5.3. Soaring to New Heights: Boeing

Boeing has embraced VR for improving assembly line processes and training. Through AR (Augmented Reality) applications, Boeing has reduced production time by 25% and cut wiring production errors almost to zero. These AR applications display step-by-step assembly instructions within the technicians' field of view, reducing reliance on complex schematics and increasing work efficiency.

For training, Boeing has developed a Virtual Flight Simulator. This application immerses pilots in a detailed 3D model of the cockpit, allowing them to familiarize themselves with the aircraft's controls without the substantial costs associated with actual flight simulators. It also enables trainees to confront potentially dangerous situations in a safe environment.

5.4. Pioneering Progress: Lockheed Martin

Lockheed Martin leverages VR and AR for spacecraft manufacturing and mission planning. The Collaborative Human Immersive Laboratory (CHIL) at Lockheed Martin uses VR to spot design errors, develop assembly strategies, and train technicians before a single physical component is manufactured. This preemptive analysis significantly reduces the time and cost of error correction usually associated with traditional manufacturing.

Further, to plan NASA's Orion mission to Mars, Lockheed Martin is using VR to simulate the mission. This ensures the spacecraft design is robust enough to withstand the rigors of an interplanetary journey, and the astronauts are aptly trained for the tasks.

The adoption of VR in aerospace industries has already resulted in considerable innovation and cost savings. As these industries tap further into the potential of VR, we can anticipate even greater applications for training, collaboration, and system design in the future. Thus, VR holds promising potential in contributing to the future of space exploration and aerospace industries at large.

Chapter 6. Advancements in VR Software for Aerospace Applications

Advancements in VR software now establish their rightful standing at the forefront of aerospace engineering. The following sections delve into the detailed examination of these cutting-edge developments.

6.1. Integration with Engineering Software

One of the major advancements in VR software for aerospace applications is its integration with advanced engineering software like CAD (Computer-Aided Design), CAM (Computer-Aided Manufacturing), and CAE (Computer-Aided Engineering). These integrations permit engineers to sculpt designs with precision in a simulated environment, thereby eliminating potential conflicts and errors that might arise during actual manufacturing. Progressions in haptic technology additionally offer tactile feedback, adding depth to design interaction.

ANSYS VRXPERIENCE, a software developed by Ansys Inc., converges simulation, VR, and physics into a single platform, allowing engineers to visualize their designs before implementing them in the physical world. As the aerospace industry leans increasingly into digitization, this sort of platform acts as a catalyst to the design, testing, and manufacturing process.

6.2. Realistic Simulations

Advancements have also been made in rendering hyperrealistic simulations to mimic real-world conditions. Software like WorldViz's Vizard implements strategic lighting, accurate physics, and fine-tuned sound systems to provide realistic spaces for design experimentation and immersive training.

This immersion seamlessly bridges understanding gaps arising due to the restrained real-world testing conditions. Programs emulate microgravity, extreme temperatures, radiation, and vacuum as though personnel are witnessing it firsthand. This prepares them for the rigors of space travel, particularly in crisis situations where quick thinking becomes a survival tool.

6.3. Training Simulators

The use of VR in training simulators has been prevalent for several decades, but recent years have seen a significant increase in sophistication. High-fidelity simulators, like BISim's VBS Blue IG, effectively mimic instrument behavior and model terrain, weather, and other environmental variables. These simulators also allow multiple users to engage in cooperative training tasks in the simulative environment, fostering team coordination.

6.4. Enhancing R&D through VR

Another considerable advancement has been made in utilizing VR for prototyping and Research & Development (R&D) within the aerospace industry. This approach reduces costs, mitigates risks, and shortens the development cycle, offering a significant competitive advantage.

Tools such as the 'Unreal Engine' create a platform for robust simulations and virtual prototyping. Engineers can test numerous design iterations and conduct failure analyses without the physical prototype's costs and risks. This iterative and agile approach facilitates optimization while boosting reliability and safety standards.

6.5. Enhanced Collaboration

The infusion of VR in collaborative undertakings is another remarkable contribution to aerospace development. Tools like Autodesk's BIM 360° Glue offer a shared virtual environment where teams from different departments or geographical locations can assess designs collaboratively. This tackles interoperability barriers, enhancing comprehension and minimizing miscommunications that could impede progress or amplify costs.

6.6. Future Prospects

The future of aerospace engineering will likely lean more heavily on VR. The advent of technologies like 5G is expected to facilitate real-time remote interactivity within virtual platforms, where tactile haptic sensations can be experienced authentically.

Quantum computing is also poised to play a crucial role by exponentially scaling computational power, facilitating more complex simulations rendered within realistic timelines.

In the grand scheme of aerospace engineering, VR has come up as the keystone for the industry's continued growth. Its capabilities to simulate, collaborate, and enhance are irrefutably vital in the increasingly complex voyage towards the unknown - towards space. The advancements in VR software for aerospace applications are equivalent to leaps for the sector, promising safer, more efficient, and more informed developments for human space travel.

Despite the perceived barriers, including cost and skepticism towards disruptive technologies, the aerospace industry's future is nearly impossible to envision without the integration of advanced VR systems. The software advances are gradually but surely redefining the scope and methodologies of the domain, taking it a notch higher towards realizing dreams of life beyond our blue planet. Ensuring design efficiency, undertaking convincing simulations, augmenting resource conservation, and reinforcing crew welfare, VR embodies the robustness required to deal with the challenges faced by aerospace engineering today.

Chapter 7. Safety and Efficiency: The Twin Benefits of VR in Aerospace

Virtual reality (VR) technologies come into play within the aerospace industry during training, designing, and even the maintenance phases. It provides an unparalleled immersive experience that serves the dual purpose of enhancing safety as well as augmenting efficiency.

7.1. The Immersive Training Ground

Virtual reality is revolutionizing the training and preparation that future astronauts undergo. Traditionally, astronauts are subjected to strenuous physical and mental training, equipped to handle the harsh realities of space, like temperature extremes, vacuum environments, and zero gravity. But there is a limit to how much the hostile environment of space can be replicated on Earth.

Enter VR, serving as the conduit that bridges this gap. Using sophisticated VR modules, astronauts can now train in a high fidelity, realistic simulation of space missions. By wearing a VR headset, they are transferred into the heart of a spaceship, where they are able to conduct complex tasks, manoeuvre machinery, and handle emergencies — all within a safe zone.

Moreover, this methodology reduces operational costs significantly. Recreating physical modules for every training situation is highly resource and cost intensive. With VR, physical resources needed for training are reduced drastically. The resultant cost savings can be channelled into other crucial aspects of space exploration.

7.2. Precise Prototyping, Seamless Designing

In the domain of spacecraft design and development, precision is of the utmost importance. Every fraction of a millimetre can change the course of a mission. In the past, the design process often involved creating physical prototypes, which were time-consuming and limited in terms of adjustments.

VR provides an effective alternative by introducing the capability for creating digital mock-ups. Designers equipped with VR devices can walk through the spacecraft's design, checking intricate details and making minute adjustments as necessary in a three-dimensional space. The margin for error is substantially reduced when one can visually inspect each detail in a virtual setting right from the design phase. This approach amplifies efficiency, speeds up development times, and aids in the creation of crafted-with-precision spacecrafts.

7.3. Enhancing Maintenance and Repairs

Maintenance and repair of aircraft and spacecraft have traditionally been a high-risk and high-cost process. It requires technicians to physically inspect and repair vast, complex machinery, not to mention the time restrictions and safety hazards associated with this method.

With VR, technicians can now receive their training in a risk-free environment, making them adept at performing the necessary tasks without the risk of damaging expensive equipment. This is especially useful when training for repair procedures on spacecraft because the cost of errors is inexpressibly high.

VR doesn't stop there! It continues to help, beyond training, during actual repair processes. For instance, if a technician is repairing a spacecraft while it is still in orbit, they can use VR to see a detailed, three-dimensional image of the spacecraft. This information can guide their repairs, ensuring that they are as accurate and efficient as possible.

7.4. Mitigating Spatial Disorientation

One of the most challenging aspects of space travel is the uncanny feeling of disorientation that astronauts often encounter. In the absence of a noticeable up or down, left or right can become confusing, leading to mishaps and hazardous situations.

VR helps in preparing astronauts for such scenarios, allowing them to train inclusively, acclimating to disorientation before they set foot in actual space. Regular exposure to a VR simulated space environment can condition their senses to adapt faster, making them less susceptible to spatial disorientation.

7.5. Conclusion: The Sum of the Parts

The application of VR in the aerospace sector is a prime example of how technology can play a transformative role in enhancing safety and efficiency. Every virtual step taken in the training ground, every digital prototype, every repair conducted with the help of VR not only brings down costs and increases the efficiency of processes within the aerospace industry but also significantly enhances the safety of the humans who venture into space.

The potential offered by VR is enormous, and the current implementations within the aerospace industry are just scratching the surface. As this symbiotic relationship continues to evolve, it promises a future where the barriers between the technological and physical worlds are less distinct, and the impossible is made achievable. As it stands, we have embarked on a thrilling journey that will redefine limits and unleash untapped potential. Time will tell what other marvels this partnership will unfold.

Chapter 8. Overcoming Challenges in Integrating Aerospace and VR

The coupling of Aerospace Engineering and Virtual Reality (VR) represents a remarkable synergy that propels both industries into uncharted territories. However, harnessing this synergy is far from seamless and has certain challenges that need to be overcome.

8.1. Aerospace Considerations

Fundamental to understanding the integration of Aerospace and VR is a rudimentary comprehension of aerospace architecture. Aerospace engineering adheres to laws of physics that are not only understood but are also inflexible and unforgiving. Error rates acceptable in other fields can't be tolerated in this space, as these can lead to cataclysmic failures.

Moreover, designing and testing aerospace systems requires a combination of high-fidelity simulators, costly prototypes, and rare physical trial runs. These processes are laborious, time-intensive, and financially draining. Idealizing such complex, high-stakes systems in a virtual environment without compromising on accuracy is a formidable challenge.

8.2. VR Environment Challenges

Virtual reality, like any other technology, has its share of challenges. VR environments right now are designed primarily for gaming or simple industrial applications. Hence, they are often incapable of providing the high-precision simulations necessary for aerospace applications.

A key problem is the difficulty of achieving real-time, physic-based rendering of complex digital models. This hampers the ability to conduct meaningful virtual tests on aerospace systems or components, as discrepancies between virtual and physical properties might lead to erroneous conclusions.

8.3. Lags and Latency issues

Latency is the time taken by a system to react to a given input. In the context of VR, high latency can invoke a dissonance between a user's actions and system's response, often leading to motion sickness or disorientation. For applications in the domain of aerospace, where swift and accurate responses are critical, latency can have grave implications.

8.4. Integration Hurdles

Unifying the worlds of Aerospace and VR calls for extensive collaboration between different sets of professionals, such as developers, designers, aerospace engineers, and VR experts. The varying technological lexicons could lead to communication bottlenecks. Achieving efficient coordination amidst such diversity is an uphill task.

8.5. Technological Constraints

While VR technologies have made remarkable progress, resolutions and refresh rates of VR displays are still far from those of human vision. Achieving this level of fidelity is particularly crucial for high-precision fields like aerospace engineering where minute details could significantly impact outcomes.

8.6. Effective Training Solutions

Among the more promising uses of VR in aerospace is the option for immersive, hands-on training for astronauts and engineers. However, it requires an ultrarealistic depiction of the space environment, which requires immense computational power and detailed data sets.

8.7. Overcoming these Challenges

Despite these challenges, the integration of Aerospace and VR is achievable. It will require constant improvements and regular iterations to VR technology to establish high-fidelity and high-precision environments. Simultaneously, employing interdisciplinary teams can ensure a cohesive and unified vocabulary for a seamless coupling of the two domains.

Additionally, advancements in computer hardware provide promising ground for the deft handling of high data processing demands. Picture distributed cloud systems; capable of multi-tasking at untold rates, delivering advanced, detailed simulations in real-time.

Further research in the field of photonics could lead to the development of VR displays with human-like resolution. Real-time, photonics-based rendering techniques could offer the path towards latency-free and high-resolution VR environments.

The uptake of technologies like machine learning and artificial intelligence could lead to more precise and specific interpretations of VR simulations, therefore aiding aerospace decision-making processes.

8.8. The Next Steps ahead

Challenges in the integration of Aerospace and VR should be seen as opportunities of growth. Each hurdle denotes a missing piece in our understanding that, once unveiled, brings us closer to the comprehensive union of the two domains. The road isn't easy, but the rewards can be transformative.

Through persistent effort and innovative solutions, we can look forward to a future where these two technologies harmoniously intertwine. It's a vision worth striving for, one that has the potential to illuminate a new path for the future of human exploration and understanding.

At the heart of it all, the relentless pursuit of knowledge and a profound sense of curiosity shape the development and progress. Bound by these common threads, the integration of Aerospace and VR will continue to disrupt the realm of possibilities, carving a path towards an unprecedented future.

This journey provides an exciting opportunity to witness and experience the remarkable confluence of advanced technologies, and their potential to reshape the world as we know it. So, buckle up for a fascinating ride as we continue to traverse the vast frontiers of aerospace technology and virtual reality.

Chapter 9. The Future of VR in Aerospace: Trends and Predictions

Delving into the spectacles of tomorrow can often be a daunting task. We are navigating uncharted territories, filled with hypotheses and predictions, yet we do have some key indicators guiding us. In the world of aerospace and virtual reality (VR), these signposts help us sketch an illustrative picture of the future. Harnessing these advancements, we can draft predictions based on current industry trends, technological progress, and research initiatives.

9.1. The Proliferation of VR Simulations

The use of VR within the realm of aerospace is notably increasing. Mirroring the trajectory of VR in other industries, it's estimated that VR simulations in aerospace will experience a growth trend. This estimate is well supported by the influx of investment into VR technologies, the number of patents being filed, and the expanded use cases across the sector.

Through these simulations, mission planning, design testing, and astronaut training are all enhanced. Traditional methods of prototyping, for instance, are expensive and time-consuming. By using VR, aerospace companies can virtually construct, deconstruct and interact with entire aircraft or spacecraft designs without spending vast amounts of resources. Such advantages are predicted to drive the growth of VR simulations in aerospace engineering.

9.2. The Advent of Haptic Feedback

Haptic technology, or tactile feedback tech, is becoming an essential element in the evolving interface between humans and computers. In VR, haptic feedback can recreate the sense of touch by applying forces, vibrations, or motions to the user. This tech allows users to experience real-world physical interactions in a virtual environment.

Future VR systems are predicted to amalgamate more deeply with haptic technologies, offering more precision and real-world simulation. Advances in this field might eventually facilitate astronauts' gloves that can simulate the feel of Martian soil or a surgeon's hand feeling a human organ remotely. The onset of such detailed haptic rendering will undoubtedly revolutionize astronaut training methods and space explorations.

9.3. Adoption of Augmented Reality (AR)

Augmented reality (AR), although tactically different from VR, shares a common endeavor: to blur the lines between the physical and digital landscapes. As opposed to completely fabricated digital environments, AR overlays virtual objects onto the real world. This technology is likely to find multiple applications in future aerospace engineering processes.

In the prediction of future trends, we foresee an extensive adoption of AR in the servicing of space equipment, both on Earth and in space. For instance, AR could guide astronauts during complex equipment repairs or prop up procedural checklists during space walks. Back on Earth, it could help technicians with aircraft maintenance or provide real-time enhancement of manual tasks.

9.4. Seamless Integration of AI

No futuristic technology outlook would be complete without pondering the role of artificial intelligence (AI) in the narrative. With AI, VR could become much more responsive and personalized. AI could learn from and adapt to individual astronauts' learning styles, speed, and aptitude, making training programs much more effective. Therefore, it's expected that AI will play an increasingly significant role in the VR-dependent future of aerospace engineering.

9.5. Accessibility and Democratization

With growing fields like Space Tourism entering the mainstream, it's crucial to consider the democratization of these technologies. As the production of VR devices increases and the costs decrease, space exploration will no longer purely be the domain of elite astronauts.

Everyday enthusiasts might soon utilize the same simulations used by astronauts in training, giving them a taste of space from their living rooms. As VR software becomes easier to develop, we might also see a surge in user-generated content, virtually opening up the Milky Way for all.

9.6. Conclusion: A Confluence of Realities

The future of VR in aerospace engineering is promising and transformative. Integrating VR, AR, haptic feedback, AI, and more advanced technologies can result in breakthroughs that can revolutionize the whole realm of space exploration. With such a convergence of realities, we will not just gain significant improvements in the traditional aspects of aerospace but will also redefine our celestial horizons. This future, although defined by cutting-edge technologies, will ultimately serve to deepen our connection with the cosmos, igniting a renewed era of exploration and discovery.

In an expounding summary, we are on the cusp of an exciting new reality that blends traditional human experience with digital sophistication, paving the way for not just space exploration, but the exploration of the endless possibilities of human endeavour.

Chapter 10. Implications on Space Travel: A New Dawn of Exploration

Before diving into the manifold implications of advancements in aerospace engineering and virtual reality on space travel, it's crucial to understand the challenges that current and future space missions have to contend with. Prominent among these are the long duration of space voyages, the extreme environmental conditions, the exposure to cosmic radiation, and the physiological and psychological effects of prolonged confinement.

10.1. The Impact of Virtual Reality Training

One of the significant ways aerospace engineering is leveraging virtual reality is in the realm of astronaut training. Gone are the days when space agencies had to build full-scale physical mock-ups of spacecraft to simulate the conditions of space travel. Today, with the advent of powerful software and high-resolution screens, astronauts can get an immersive and realistic training experience without ever outstepping the confines of a VR room.

Astronauts can don VR headsets and be instantly transported into a virtual space environment. They can experience, in minute detail, the landscape of a spaceship or an extraterrestrial surface. This isn't just cost-effective but also allows for a wider range of simulations. A malfunction in an oxygen tank, a meteor strike, or a docking procedure can all be practised to perfection within the safe confines of a VR room.

10.2. Overcoming Isolation and Confinement

Beyond just training, virtual reality holds transformative solutions for one of the most potent psychological challenges of space travel: isolation from the earth. Astronauts, stuck in the limited confines of a spacecraft on a journey that could last months or even years, are prone to feelings of loneliness and depression.

Virtual reality can play a critical role in alleviating this psychological pressure. Through VR, astronauts can 'return' to earth, stroll through a forest or walk alongside a beach, engaging senses and providing a semblance of normalcy in an otherwise alien environment. This is of paramount importance when one recalls the impact of psychological well-being on overall mission success.

10.3. On-Demand Medical Assistance

With advancements in virtual reality, prompt and convenient medical assistance becomes a reality even in outer space. A medical officer can guide an astronaut in performing a complex medical procedure through virtual reality. Simultaneously, an AI-backed system could provide real-time feedback, ensuring that astronauts' health never takes a backseat.

10.4. Real-time Repair and Maintenance

Another promising application of virtual reality in space travel is in equipment repair and maintenance. Over the course of a voyage, the spacecraft's multiple systems might need repairs or check-ups. A VR system, integrated with the spacecraft's technical manuals, could guide an astronaut visually in performing a repair, significantly lowering the chances of error.

10.5. Planning and Testing Mission Strategies

Virtual reality could also be a powerful tool for space mission strategists. Simulating an entire space mission in VR allows scientists and engineers to identify potential issues, optimize strategies, and experiment with different scenarios to ensure a successful mission.

Virtual reality, paired with aerospace engineering, is set to usher in a new era of space exploration. From efficient training and mental health management to on-demand medical assistance and real-time repair and maintenance, the partnership promises to revolutionize the way we approach and experience space travel. It lays the groundwork for a future wherein our ambitions of exploring the cosmos are no longer shackled by the physical and psychological limitations that have held us back thus far. With the complexity of space travel broken down into virtual bytes, we are on the cusp of a new dawn in exploration. From here, the only way to go is up. And with virtual reality along for the ride, the journey promises to be nothing short of transformational.

Chapter 11. Embracing VR in Aerospace: A Strategic Guide

Aeronautics has always harnessed the latest technology to go a step further, break boundaries and redefine limits. Today we are witnessing the exciting confluence where the space arena and digital reality blur the lines between the tangible and virtual. This convergence carries an intrinsic potential to revolutionize both the aerospace sector and virtual reality technology. It promises not just to add another dimension to our understanding of the universe, but to transform our approach to space exploration entirely.

11.1. The Need for VR in Aerospace

Development and testing in aerospace sector come with significant cost, risk, and time investment. Traditionally, testing of spacecraft or aircraft designs involves the construction of physical prototypes. These prototypes are tested extensively against simulated conditions representing space or flight environments. However, this process is resource intensive, demands comprehensive synchronization of various factors, and leaves a broad margin for error.

Virtual reality offers an efficient solution to these challenges. With VR, engineers can simulate a space environment without the need of physical prototypes or setup. The engineers can interact with the virtual models and simulate situations, thus reducing cost, saving time, and bringing down the risk significantly.

11.2. VR in Aerospace Design and Training

Designing spacecraft, planes, and satellite equipment is a complex process. Engineers and designers can leverage VR to visualize, interact, and modify their design prototypes in a virtual environment before it transitions from design to production. This immediately eases the iterative process of design and development.

VR is also transformative in the area of pilot training and astronaut training programs. Trainees can immerse themselves in accurate, often dire, scenarios without actual risk – helping them prepare adaptively for critical situations without expensive, logistically complicated, and potentially hazardous physical drills.

11.3. Precision & Accuracy through VR

VR can simulate the intricate behavior and demands of the extreme conditions in aerospace missions with an astonishing degree of accuracy. It enables engineers to test and optimize against a wide range of variables, such as gravitational forces, thrust power, fuel consumption, heat shield functionality, and more.

In addition to testing, VR can facilitate highly precise assembly and alignment regimes for space vehicles, aircrafts, and satellites. This will result in mitigating mission-critical failures in spacecraft functions, satellite positioning, or aircraft flight stability.

11.4. VR for Mission Simulation and Exploration

One of the more ambitious areas of VR in aerospace is the use of the technology for mission simulation and space exploration. By transforming complex data into interactive 3D simulations, VR can replicate microgravity environments or extraterrestrial landscapes to enable astronauts to prepare for actual conditions encountered in space or other planets.

NASA's Hybrid Reality Lab uses a blend of physical and virtual models to train astronauts in repairing space equipment. In another pioneering move, private company SpaceX's spacecraft Crew Dragon was completely designed using VR technology, making it the first spacecraft where simulation in VR preceded actual flight.

11.5. The Future Awaits

The possibilities of VR in aerospace are far from fully realized. They promise a future where astronauts navigate foreign terrains virtually before landing, where space probes are driven remotely in real-time with the same dexterity as navigating through your own home, and where everyone can experience the thrill of space travel in the safety of their living room.

However, such exciting prospects do come with their share of challenges. Issues such as latency in data transmission, the fine balance between realism and computational load, making the VR experience more intuitive and comfortable, and developing regulatory frameworks are some areas that call for innovation and R&D.

11.6. Conclusion

As we stand on the cusp of one of the most riveting chapters of technological advancement, it is clear that the blend of aerospace and VR introduces a new era of exploration and understanding. It is the key that allows us to continue pushing beyond our known boundaries and make the impossible, possible. The journey might be laden with challenges, but the rewards are colossal. Embracing VR in aerospace not only signifies a strategic forward move but should be viewed as a guided leap towards the future.

As the virtual and real continue to converge and blur their defined boundaries, a new cosmos awaits our exploration, right from our earthly abode, and VR will be our cosmic vehicle. The journey, indeed, is just beginning.